Full STEAM Ahead!

Engineering Everywhere

Ideas from Nature

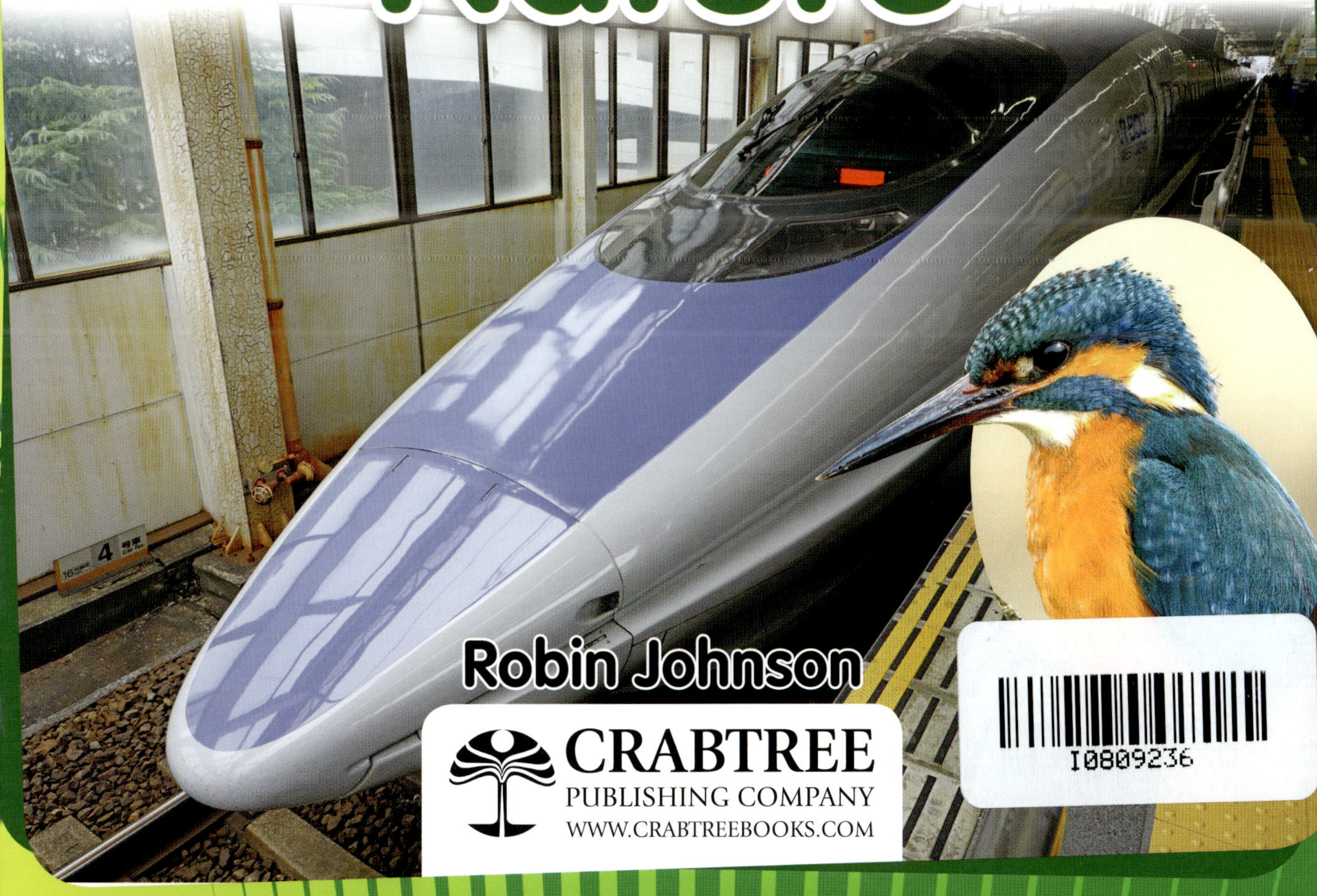

Robin Johnson

CRABTREE
PUBLISHING COMPANY
WWW.CRABTREEBOOKS.COM

Title-Specific Learning Objectives:

Readers will:

- Explain how problems are solved in nature.
- Describe how nature inspires solutions to human problems.
- Make connections between ideas in a text.

High-frequency words (grade one)	**Academic vocabulary**
are, can, have, make, that, they, this, who	adaptations, design, energy, engineers, environment, repel, reservoir

Before, During, and After Reading Prompts:

Activate Prior Knowledge and Make Predictions:
Hold up a picture of a turtle and have children describe the animal, naming its parts. Point to the shell and ask what its purpose is. Children should understand that the shell protects the soft body of the turtle. Now ask them if they can think of things people have created for protection that remind them of a turtle's shell. Children might mention body armor, a football helmet, or even a tank.

Tell children that they will be reading a book called *Ideas from Nature*. Ask them what they think the title means, and what they think the book will be about.

During Reading:
After reading pages 16–17, ask children to look again at the pictures. Ask:

- In what ways are the duck's feet similar to the flippers and paddles?
- In what ways are they different? Why do you think the inventors made these differences?

After Reading:
Ask children to make a T-chart. On one side, have them list the things from nature that were pictured in the book. On the other side, they should list the human-made inventions that engineers created based on these natural objects.

Author: Robin Johnson

Series Development: Reagan Miller

Editors: Bonnie Dobkin, Janine Deschenes

Proofreader: Melissa Boyce

STEAM Notes for Educators: Bonnie Dobkin

Guided Reading Leveling: Publishing Solutions Group

Cover, Interior Design, and Prepress: Samara Parent

Photo research: Robin Johnson and Samara Parent

Production coordinator: Katherine Berti

Photographs:
iStock: kali9: p. 17 (b)
Shutterstock: EQRoy: title page (l)

All other photographs by Shutterstock

Library and Archives Canada Cataloguing in Publication

Title: Ideas from nature / Robin Johnson.
Names: Johnson, Robin (Robin R.), author.
Description: Series statement: Full STEAM ahead! | Includes index.
Identifiers: Canadiana (print) 20200165879 |
Canadiana (ebook) 20200165887 |
ISBN 9780778771814 (hardcover) | ISBN 9780778772668 (softcover) |
ISBN 9781427124593 (HTML)
Subjects: LCSH: Biomimicry—Juvenile literature. |
LCSH: Engineering—Juvenile literature. | LCSH:
Engineers—Juvenile literature. | LCSH: Technological innovations—
Juvenile literature.
Classification: LCC TA164 .J65 2020 | DDC j620—dc23

Library of Congress Cataloging-in-Publication Data

CIP available at the Library of Congress
LCCN: 2019058887

Printed in the U.S.A./032020/CG20200127

Table of Contents

Crabtree Publishing Company
www.crabtreebooks.com 1-800-387-7650

In Canada: We acknowledge the financial support of the Government of Canada through the Canada Book Fund for our publishing activities.

Published in Canada
Crabtree Publishing
616 Welland Ave.
St. Catharines, Ontario
L2M 5V6

Published in the United States
Crabtree Publishing
PMB 59051
350 Fifth Avenue, 59th Floor
New York, New York 10118

Published in the United Kingdom
Crabtree Publishing
Maritime House
Basin Road North, Hove
BN41 1WR

Published in Australia
Crabtree Publishing
Unit 3 – 5 Currumbin Court
Capalaba
QLD 4157

Engineers Solve Problems

Engineers are people who use math, science, and **creative thinking** to solve problems. They **design** things that help people meet their needs.

Engineers design tools that let people study nature up close.

Engineers come up with ideas that make our lives easier. These shoes don't need shoelaces!

Engineers design things that meet our need to be active and have fun.

Nature Solves Problems

Nature solves problems too! Animals may change their environment to meet their needs. They build nests, dig holes, and make other changes where they live. Some animals and plants developed **adaptations** that help solve problems.

Birds design nests that fit their environment.

Spiders make strong, sticky webs to catch insects to eat. They use body parts called spinnerets to make the webs.

Cactuses grow in places that do not get much rain. Their thick stems are an adaptation that helps them store water.

Solutions from Nature

Engineers can study the natural world to find solutions to human problems. They look at how animals and plants live, move, get food, and stay safe. Engineers use what they learn to design new **technologies**.

Some engineers study the wings of birds. This helps them design better airplane wings.

Kangaroo mothers carry their babies in pouches on the front of their bodies. Engineers have designed technologies that let people carry their babies the same way.

Sun Power

Plants get **energy** from the Sun. They use sunlight, water, and air to make their own food. Engineers design technologies that let people use the Sun's energy too.

The leaves on sunflowers and all other plants take in energy from the Sun.

Engineers design **panels** that collect and store energy from the Sun. The panels work like leaves on a plant.

These students are using a model to learn how people use the Sun's energy to heat and power their homes.

Sticky Problems

Plants and animals can give engineers ideas for all kinds of technologies. Some technologies solve the problem of how to make one thing stick to another.

Have you ever had burrs stick to your clothes? These prickly parts of plants gave engineers an idea for a type of fastener.

The tiny hairs on a gecko's feet help it climb up almost anything.

Engineers used what they learned from geckos to make better climbing shoes.

Staying Dry

Some plants have leaves that **repel** water. Water rolls off the leaves so the plants stay dry. Engineers study these plants to design new **materials**. These materials repel water and help keep people dry.

The water lily plant grows in water. It does not need to collect water on its leaves, like other plants do. Its leaves repel the water.

Umbrellas, rain boots, and raincoats meet people's need to stay dry when it rains.

Webbed for Water

Some animals have webbed feet for swimming. They use their feet like paddles to move through the water. Engineers design technologies that help people move through water too.

Ducks and some other water birds have webbed feet. They have skin between their toes that helps pull them through water.

Engineers design flippers to help people swim. Flippers are flat rubber shoes that look like webbed feet.

These girls are learning how to use paddles to move through the water.

Busy Beavers

Beavers are busy animal engineers! They build dams in streams to block the flow of water and create ponds, where they build their homes. Engineers build dams too. They use the dams to control the flow of water in towns and cities.

Beavers use tree branches, mud, and rocks to build dams. They really change their environment to meet their needs!

Engineers use dams to create lakes called reservoirs. A reservoir stores fresh water for people to use. Dams also help keep rivers from flooding.

Need for Speed

Cheetahs are the fastest land animals on Earth. An engineer studied cheetahs to learn how they run and jump. Then he designed **artificial** legs for people. We have nature to thank for many inventions like these!

A cheetah is built for speed! It has a light body and long, powerful legs.

This type of artificial leg allows people to run fast.

Words to Know

adaptations [ad-*uh* p-TEY-sh*uh* ns] noun Changes in a plant or animal that helps it survive in its environment

artificial [ahr-tuh-FISH-uhl] adjective Made by humans

creative thinking [kree-EY-tiv THING-king] noun Using your mind to make up new and original ideas

design [dih-ZAHYN] verb To make a plan for how something is made or built

energy [EN-er-jee] noun The power to do work

materials [m*uh*-TEER-ee-*uh*ls] noun Things from which something is made

panels [PAN-ls] noun Thin, flat pieces

technologies [tek-NOL-*uh*-jees] noun Things people create to make life easier, safer, and more fun

repel [ri-PEL] verb To push away

A noun is a person, place, or thing.

A verb is an action word that tells you what someone or something does.

An adjective is a word that tells you what something is like.

Index

About the Author

Robin Johnson is a freelance author and editor who has written more than 80 children's books. When she isn't working, Robin builds castles in the sky with her engineer husband and their two best creations—sons Jeremy and Drew.

To explore and learn more, enter the code at the Crabtree Plus website below.

www.crabtreeplus.com/fullsteamahead

Your code is:
fsa20

STEAM Notes for Educators

Full STEAM Ahead is a literacy series that helps readers build vocabulary, fluency, and comprehension while learning about big ideas in STEAM subjects. *Ideas from Nature* helps readers understand how to make connections between the ideas they read about in a text. The STEAM activity below helps readers extend the ideas in the book to build their skills in science, technology, and art.

Inventions from Nature

Children will be able to:

- Analyze an image.
- Brainstorm inventions based on the images.
- Design an invention.

Materials

- Animal Experts Handout
- Invent It! Worksheet
- Art supplies for drawing

Guiding Prompts

After reading *Ideas from Nature*, ask children:

- What is an engineer?
- How do engineers get ideas from nature?
- What are some examples of ideas that came from nature?

Activity Prompts

Review the definition of an engineer with children. (Engineers are people who use math, science, and creative thinking to solve problems and meet needs.) Tell children that they will have the chance to think like an engineer. Remind children of the activity they did before reading the book, when they figured out what inventions were inspired by a turtle's shell. Tell them that their goal is to come up with an idea or invention based on something in nature.

Divide children in groups of three or four. Provide each group with an Animal Experts Handout and an Invent It! Worksheet. Tell children that if they study these animal experts carefully, they can get ideas for new inventions, just like other engineers got ideas from the turtle's shell.

Have children choose an animal from the handout. Direct them to go through the steps of the worksheet with their group. Once they have come up with their idea, allow each child time to draw it on a separate piece of paper. Encourage group members to compare their final pictures. In what ways were the pictures the same? How were they different?

Extensions

- Have each group practice presentation skills by explaining their finished invention to the rest of the class. Make sure they cover the following points, which can be written on the board: What is the invention? What does the invention do? How did an animal inspire the invention?

To view and download the worksheet, visit **www.crabtreebooks.com/resources/printables** or **www.crabtreeplus.com/fullsteamahead** and enter the code **fsa20**.